WHAT DOES IT LOOK LIKE?

IT HAS A SHELL!

BY KATHLEEN CONNORS

Gareth Stevens PUBLISHING

first concepts

Shells keep an animal's body safe.

Shells help animals hide.

Many animals with
shells live in water.
Sea turtles have a shell.

Clams have a shell.

Crabs have a shell.

Lobsters have a shell.

Oysters have a shell.

Some animals with shells live on land. Snails have a shell.

Tortoises have a shell.

Armadillos have a shell.

Turtles live on land and in water.

Turtles have a shell.

Which animal has a shell?

Please visit our website, www.garethstevens.com. For a free color catalog of all our high-quality books, call toll free 1-800-542-2595 or fax 1-877-542-2596.

Cataloging-in-Publication Data

Names: Connors, Kathleen.
Title: It has a shell! / Kathleen Connors.
Description: Buffalo, NY : Gareth Stevens Publishing, 2026. | Series: What does it look like?
Identifiers: ISBN 9781482470819 (pbk.) | ISBN 9781482470826 (library bound) | ISBN 9781482470833 (ebook)
Subjects: LCSH: Shells–Juvenile literature. | Mollusks–Juvenile literature. | Body covering (Anatomy)--Juvenile literature.
Classification: LCC QL405.2 C747 2026 | DDC 594–dc2

Published in 2026 by
Gareth Stevens Publishing
2544 Clinton Street
Buffalo, NY 14224

Designer: Andrea Davison-Bartolotta
Editor: Kristen Nelson

Photo credits: Cover, p. 1 (tortoise) SUPIDA KHEMAWAN/Shutterstock.com; cover, p. 1 (background) briddy/shutterstock.com; p. 3 (armadillo) Chase D'animulls/Shutterstock.com; p. 3 (crab) Kmikhidov/Shutterstock.com; p. 3 (snail) Wisista C/Shutterstock.com; p. 3 (tortoise) FOTOGRIN/Shutterstock.com; p. 5 Sakis Lazarides/Shutterstock.com; p. 7 Bence Sibalin/Shutterstock.com; p. 9 Jesus Cobaleda/Shutterstock.com; p. 11 Rostislav Ageev/Shutterstock.com; p. 13 Oleg Kovtun/iStockphoto.com; p. 15 Ivan Marjanovic/Shutterstock.com; p. 17 Nick Greaves/Shutterstock.com; p. 19 Christopher Biggs/Shutterstock.com; p. 21 Leandro Inoe Coelho/Shutterstock.com; p. 23 (bird) New Africa/Shutterstock.com; p. 23 (gecko) Robert Eastman/Shutterstock.com; p. 23 (lion) Eric Isselee/Shutterstock.com; p. 23 (sea turtle) Mr.wutthiphat vimuktanont/Shutterstock.com.

Printed in China

CPSIA compliance information: Batch #QSGS26: For further information contact Gareth Stevens at 1-800-542-2595.